Hand milking in a Cardiganshire farmyard, 1897.

DAIRYING BYGONES

Arthur Ingram

Shire Publications Ltd

CONTENTS

The dairy in the making	3
Milk	5
Buttermaking and clotted cream	13
Cheesemaking	23
Further reading	30
Places to visit	30

Printed in Great Britain by C. I. Thomas & Sons (Haverfordwest) Ltd, Press Buildings, Merlins Bridge, Haverfordwest, Dyfed.

Copyright © 1977 and 1987 by Arthur Ingram. First published 1977; reprinted 1980, 1983. Second edition 1987. Shire Album 29. ISBN 0 85263 866 3.
All rights reserved. No part of this publication may be reproduced or transmitted in any form or by any means, electronic or mechanical, including photocopy, recording, or any information storage and retrieval system, without permission in writing from the publishers, Shire Publications Ltd, Cromwell House, Church Street, Princes Risborough, Aylesbury, Bucks, HP17 9AJ, UK.

COVER: A detail from 'The Dairy Farm' by James Ward (1769-1859), from a nineteenth-century mezzotint.

ACKNOWLEDGEMENTS
The author acknowledges the help of the following in the preparation of this book: Mr R. Fowler; Mr J. Forward; Mr G. Owen, Lackham College of Agriculture; Mr J. Bevan, Welsh Folk Museum. The photographs were taken by Malcolm Harris; those on pages 1, 13, 15 (all), 20 (top) and 22 were taken at the Welsh Folk Museum, St Fagans, Cardiff; those on pages 6 (bottom), 7, 8 (centre), 11 (right), 18 (right), and 27 at Lackham College of Agriculture, Lacock, Wiltshire; and those on pages 8 (bottom), 17, 23, and 26 at Mr John Forward's Longhedge Collection, Corsley, Wiltshire.

Hand milking by the light of a hurricane lamp on a winter evening on a Wiltshire smallholding, 1976.

Shorthorn cows from Blackwood's 'The Book of the Farm'.

THE DAIRY IN THE MAKING

It is probable that dairying in its most primitive form was practised by neolithic man as he abandoned his nomadic, hunting and gathering existence and adopted a more sedentary way of life growing crops and rearing stock on the southern uplands of Britain. Milk was a welcome addition to his diet. From the bronze age, sheep, similar in general appearance to the Soay sheep of today, were the main source: they were kept as much for their milk as for their flesh and hide, as were goats. Cattle, by virtue of their superior strength and size, presented early man with far greater problems of handling and confining. They continued for some considerable time to be hunted in a wild state for meat and hide. The Celts of the north and east pioneered their domestication and over a long period of progressive breeding in captivity cows became more docile and tractable. The volume of milk produced by each individual cow, compared to sheep and goats, undoubtedly established it as the primary milk-producing animal by the end of the sixteenth century.

Dairying in the modern sense developed gradually. Early man catered only for his more immediate needs but certainly consumed the curdled milk which provided the basis for the evolution of sour milk cheese well before Roman times. Butter-making was probably introduced by the Celts. In Anglo-Saxon times butter and cheese production was well established and during the middle ages most cottagers had a milch animal. The peasant's cow, either his own or loaned by the lord of the manor and returned when it became dry, and sometimes a goat as well, supplied his family needs. The self-sufficient household undertook all its own processing and dairying was a kitchen industry. The manorial or noble households were the first to provide special premises for milk processing to cope with a greater volume, but at this time even they were governed by the difficulties of providing fodder during the winter and slaughtered a large percentage of their stock in the autumn.

By the mid eighteenth century the situation was considerably altered. The enclosure of what had formerly been common grazing land, especially in south and midland England, had forced many landless labourers to abandon the family cow and increasingly farmers and estate owners found it profitable to specialise in producing cheese and butter for sale. New and improved fodder crops, efficiently harvested, enabled farmers to overwinter the stock on a diet which maintained the quality and quantity of their milk. And the selective breeding of dairy herds had begun.

Most farms sited the dairy next to the farmhouse on the north wall, often under the shade of trees so that it would keep cool; the roof of the building would for preference be thatched or stone-tiled to insulate against summer heat. The design of the early dairies varied considerably according to taste and finances. Many were just a single room where the milk was placed in churns, the milk buckets washed and the butter made. Others had a cheese room attached and often a cheese storage room running from that again. Paramount importance was placed on providing easily cleaned surfaces, usually of stone, and plenty of ventilation, for a good circulation of air was desirable. A few of these old dairies still survive, but most were replaced in the Victorian era. Victorian gentlemen farmers, estate owners and even many of the less wealthy small farmers showed a great enthusiasm for dairying, producing beautifully tiled, spotlessly clean dairies and a multitude of innovations to fill them. It is from this era that most of today's dairying bygones come.

Three-legged and four-legged milking stools.

Canted milking buckets produced by the Lister Company, with a three-legged milking stool.

MILK

The constituents of cow's milk are as follows: water 87.55 per cent, lactose 4.60 per cent, butterfat 3.60 per cent, casein 3.10 per cent, mineral salts 0.75 per cent, albumen 0.40 per cent. These are average figures varying from breed to breed and between individuals. Channel Island breeds, Jersey and Guernsey, are noted for the high butterfat content of their milk and are specialist dairy cows, as are the modern Friesian and Ayrshire cattle.

However, the more popular breeds in England from the eighteenth to the early twentieth century were dual-purpose cattle, bred to produce an adequate quantity and quality of milk, but also with an eye to obtaining good beef eventually. By far the most popular of these breeds was the Shorthorn. This was a medium large red-roan coloured cow with good fattening qualities, also noted for the quality of its hide. Towards the end of the nineteenth century it gradually became more specialised and was called the 'Dairy' Shorthorn. A popular dual-purpose breed in Scotland was the Galloway, a polled variety. Other noted breeds for milking in the nineteenth century were Suffolk Duns, Red Polls, Welsh Blacks, Kerry cattle from Ireland, South Devons, considered by many to be the finest of them all, and the Old Gloucester cattle whose milk was at one time used exclusively in the making of double Gloucester cheese.

Milking was normally carried out twice daily, in the early morning and late afternoon. We are now accustomed to the sight of cattle being driven to milking parlours, but in the nineteenth century the bucket was as often as not taken out to the cow in the field or in the cow yard in the winter. The milkmaid sat on her milking stool, her head buried deep into the haunches of the cow, and with a gentle stroking motion on each quarter directed a

ABOVE: *Early nineteenth-century coopered wooden handle-piece pail (left) and late nineteenth-century metal reproduction of the same design (right).*
OPPOSITE TOP: *Goat-milking stools and pail.* OPPOSITE CENTRE: *Designed as a pig-slaughtering bench, the pig gib was frequently used by cottagers to stand their goats upon, raising them to a more convenient height for milking.* OPPOSITE BOTTOM: *Combined milk bucket and seat, produced at South Marston in Wiltshire.*

stream of warm, frothing milk into her pail.

Milking stools were normally three-legged, to give them stability on uneven ground. About 12 inches in height, they were round, hexagonal, octagonal or half-octagonal with one long straight edge, depending on the favoured local design. They were carved from a solid piece of elm, normally about 1½ inches thick, into which the three legs approximately 1½ inches in diameter, normally of ash or hazel, were tenoned and secured with a wooden wedge. Rectangular stools with four legs were also used, as were stools with a single support. The method of construction was the same as for the three-legged type. In Wiltshire, where the single-legged stool was used, the other types were derisively referred to as 'lazy-legged stools'.

Milking pails and buckets varied even more widely than stools. The pail probably derived its name as a corruption of the word *pale* (a flat strip of wood), as early milking pails were constructed from panels of wood bonded with iron hoops in the manner of a barrel. They were wide and squat in appearance, with one of the panels extended some 5 inches beyond the rim of the pail and carved to form a handle. Later pails were the same in the basic manner of construction but were shaped more like modern buckets and the carved wooden handle-piece was replaced by the familiar iron loop-over handle fixed to metal swivel lugs on opposing sides of the pail.

The Victorians mass-produced metal milk buckets in several designs such as the Lister Company's simple common bucket, a reproduction of the old 'handle-piece' pail, and the canted bucket with an opening diminished in size and located in the side, the whole bucket being similar in shape to the pottery storage jars now sold for salt and flour. The idea of having the mouth canted in this manner was to

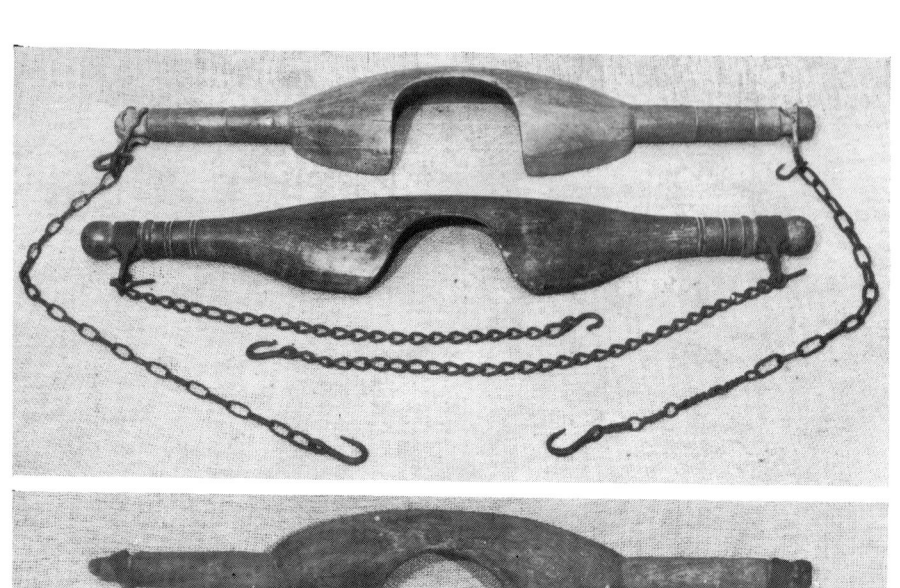
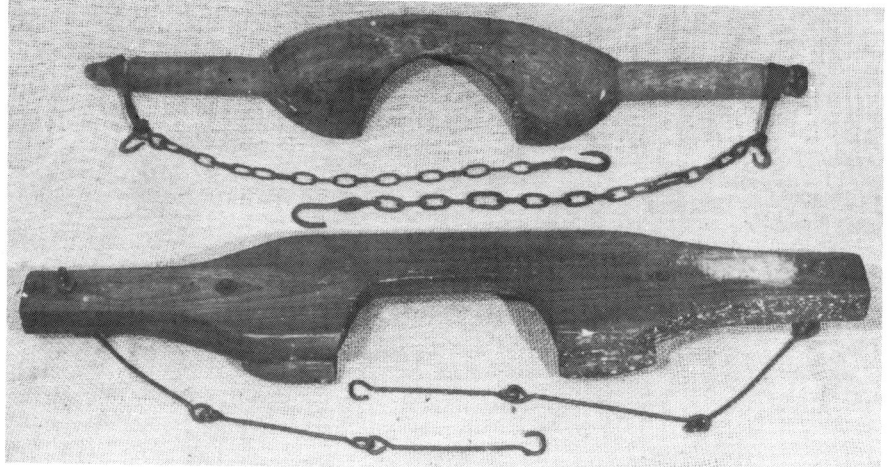
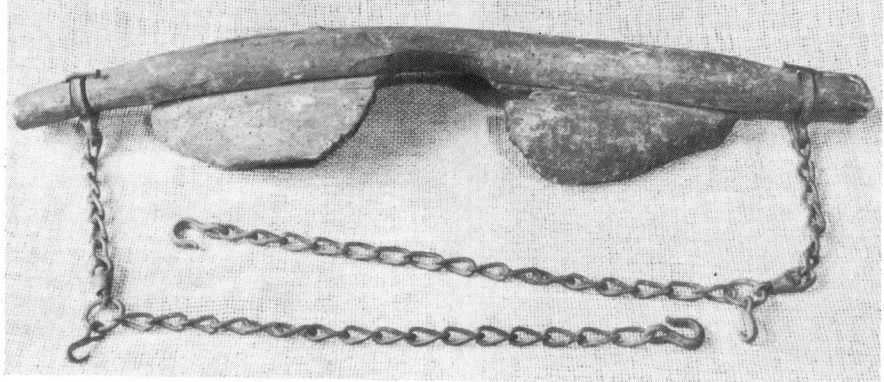

ABOVE LEFT: *Milk cooler and 17-gallon churn.* ABOVE RIGHT: *Cake-breaking machine, with a slab of linseed cattle cake.*
OPPOSITE: *(from the top) Nineteenth-century dairy yoke. Early twentieth-century yoke. Nineteenth-century yoke. Crude home-made nineteenth-century yoke. Early nineteenth-century yoke crudely fashioned from a conveniently shaped length of branchwood, nailed to roughly carved shoulder pieces.*

reduce the risk of dirt falling into the milk or the cow thrusting her hind foot into the milk, a common enough hazard with the open-topped bucket. A firm called Carlton of Manor Farm, South Marston, Wiltshire, went one step further and combined the milking stool with a bucket which lay almost horizontally. Its appearance was rather that of an elongated coal scuttle supported near its mouth by a low double support. On the upper surface, at the rear, was fixed a small, perforated metal seat. The whole thing was about 2 feet long.

Goats were often raised to a convenient height for milking by standing them on *pig gibs*, low benches used to slaughter the family pig. Alternatively the milker sat on a small, four-legged wooden stool, some 7 inches high, and milked into a tiny wooden pail about 6 inches tall and 5 inches in diameter, carved from solid wood and bound with iron.

The more orthodox-shaped buckets and pails were carried from the cow to the dairy in pairs with the aid of a shoulder yoke. This beam of wood was usually carved from willow for lightness, but ash and sycamore were also used. It was carved broad and concave at its centre to fit comfortably across the shoulders, with a recess cut in the leading edge to fit around the neck of the wearer. The ends tapered in the round to terminate some distance clear of the wearer's shoulders. From these ends hung adjustable chains with hooked ends on which the buckets were suspended. Some very crude early yokes can be found which are no more than a

ABOVE: *50-gallon milk-delivery churn mounted on a horse-drawn milk float.*
BELOW: *Milk-delivery perambulator, used around the streets of Devizes, Wiltshire, between 1880 and 1908.*

ABOVE LEFT: *Milk-delivery can and pint measure.*
ABOVE RIGHT: *Vacuum milking bucket, with double set of teat cups (1917).*

conveniently shaped stick nailed to a crudely carved centre board.

The bottling of milk was not widely adopted until after the First World War. In earlier times milk could be bought direct from a cow or goat led from door to door or from a milk man or maid carrying covered pails on a yoke. Later large tinned churns each containing about 50 gallons, were taken round by a pony and trap. The housewife brought her own jug to the churn; the milkman would fill his measure from the brass tap at the base of the churn and transfer milk to the jug. Smaller churns were carried by delivery perambulator. This was a horizontal Y-shaped frame with a wooden cross-piece handle with which it was pushed by the milkman. At the front two extremes of the Y frame were two iron spoked wheels 3 feet in diameter and two pivoted hooks from which a churn of 25 or 30 gallons capacity was suspended.

Not all the milk consumed in larger towns came from rural areas. London, in particular, was noted for its urban cowhouses built near the heart of residential areas. They consisted of milking parlour, dairy and cow yard. It was in the yard that the animals spent their days feeding on very mixed fodder, at worst left-over vegetables and at best hay and eventually concentrate food, which came in the form of large flat slabs of cotton or linseed cake. These were broken down to a palatable size with a machine called a cake mill, which was basically a pair of spiked rollers held in a cast-iron frame and turned by handle. The spiked rollers reduced the slabs to fragments, which were collected in a tray at the base.

The growth of the railway network

increased the range of rural suppliers and after 1850 milk was carried to the towns by rail. To reduce the risk of souring on the journey the milk cooler was devised. This was simply an iron stand which supported at about shoulder height a tinned tank into which the milk was poured. In the front of the tank was a brass tap which allowed the milk to flow gently on to the cooler. The cooler itself was, in effect, a corrugated water jacket. As the milk trickled down the outside of some nine or ten horizontal corrugations cold water was forced through pipes inside. The milk collected at the bottom in a shallow trough with a brass bung plug, then flowed into a 17-gallon churn which was placed beneath. The bung plug was necessary in order to stem the flow of milk whilst churns were changed.

By the turn of the century many large dairy and cheesemaking firms were established in the rural areas and replaced the railway station as the target of the morning milk rush. Many of these dairies and many of the farms had their own local milk rounds.

The replacement of hand-milking by a mechanical process was a very protracted affair which began in earnest in the mid nineteenth century. The transition was not really completed until around 1950 and in the 1970s many smallholdings which cannot afford to install an expensive milking system still use the old hand method.

In the 1850s a number of people applied their minds to the problem of speeding up milking. In 1862 Colvin, an American, produced a machine which worked on a vacuum principle. Four rubber cups were fitted to the cow's teats and the vacuum created by vigorously pumping two handles up and down extracted the milk very rapidly from the cow's udder into the integral bucket. The rigorous stress of a constant vacuum suction of this nature, however, was injurious to the animal—indeed blood was often drawn off with the milk—and the idea was scrapped. But the need was still there and in the 1880s another machine, the lactator, was tried. It was suspended beneath the cow and worked on the principle of a hand crank operating revolving belts which in turn operated a pair of adjustable rollers that gripped each teat. When the contrivance was operated the rollers gently stroked the teats much in the manner of a hand-milker's fingers. It seems an unlikely contraption to revolutionise the dairying world and it passed into obscurity. Various other attempts were made to establish vacuum machines but all failed because of the delicate nature of the cow's udder, which could not withstand the harshness of unbroken suction.

It was in 1895 at the Darlington Royal Show that the breakthrough finally came, when a Glaswegian, Dr Shields, introduced his 'Thistle' pulsating machine. Like earlier vacuum machines it still relied on cups attached to the cow's teats extracting milk by suction, but its great innovation was a trip mechanism, called the pulsator, which at regular intervals briefly punctuated the suction. This proved to be much less harmful to the cow. It was widely adopted and, although somewhat modified, is still the principle used in the milking machine of today.

A certain amount of country lore is attached to the subject of milk. For instance, a good dairy farmer would always closely inspect a fresh pasture before introducing his cattle to it, not merely to ensure there was sufficient grass, but to guard against milk-taint. Certain plants the cow might eat can radically alter the taste of milk and subsequently of the butter and cheese products. Members of the mint family, buttercups, wild garlic, ivy, fool's parsley, marsh marigolds, turnip, tansy, ox-eye daisy, oak leaves and acorns are among the better-known plants which can affect the milk in this way. Many country folk believed that the milk from a red cow surpassed all other in quality and that the beestings, the thick milk produced by a cow immediately after calving, had great health-giving properties: it was particularly prized for making superior egg custards. In Somerset and Wiltshire the old iron pump in the dairy was often wryly referred to as the cow with the iron tail, acknowledging the too common practice of watering the milk. And it was known for some West Country farmers to boil snails bruised in milk and add them to cream. It made, I am told, a most substantial improvement to its thickness.

Hand-cranked cream separator.

BUTTERMAKING AND CLOTTED CREAM

Because it took up less space and required less equipment than cheesemaking, buttermaking could be undertaken in large farmhouse kitchens as well as in the dairy itself, and it was not uncommon for cottagers to churn all their own butter.

Preparations for making butter had to begin some seventy-two hours before the actual churning. The milk to be used would be smelt to detect any taint. It was not cooled with the rest of the milk but was poured into a *setting dish,* a large shallow pan made of either earthenware or tin, and left overnight, by which time the cream had separated from the milk. This operation was known as 'setting the milk'. The cream was skimmed off by using a *skimmer* or *fleeter,* normally made of tin, which resembled a shallow, perforated saucer with a handle. The cream was then covered with muslin and kept for forty-eight hours to ripen.

By the 1890s a machine called a *separator* had been devised to replace overnight setting. Milk, warmed to aid separation, was poured into a tank at the top. It passed into a chamber fitted with a float and then through a strainer into a chamber which revolved at great speed, subjecting the milk to a centrifugal force which caused the heavier skim milk to fly

Cream setting pans and fleeter.

to the outside while the lighter cream remained near the centre. They were then channelled separately to emerge from two different pipes. These machines could be hand-cranked, horse-geared or power-driven, and all had the very high gearing necessary to create the speed required to perform the task. They were extremely efficient but also costly, so in the main they were confined to larger dairies.

The next stage was the churning, in one of the numerous designs of butter churn used through the ages. One of the earliest types was the *plunger churn,* an upright cylindrical churn tapering towards its top. It was of coopered construction, similar to a barrel, bound in iron or withy bonds. Usually about 3 feet tall, it was fitted with a wooden lid held in position with iron clasps. From an aperture in the lid protruded the handle of the *agitator,* a 'broom' handle which was fitted at the bottom with a perforated wooden disc. The churn was partly filled with cream and the agitator was dashed up and down inside the churn until the butter eventually formed, hence the alternative names of *dash churn,* and in some localities *plump churn* or *knocker churn.* This method was first used as early as the sixteenth century when it replaced churning butter by shaking cream vigorously in a skin, a method used since time immemorial.

The *rocker churn* was also an early development. This was an oblong hollow wooden container, with rounded ends, hanging on a double-X piece wooden frame, rather like a swing boat in a fairground. A small handle at either end enabled the operator to sit beside the churn and keep the motion going with the occasional gentle push. The butter was formed by the cream swilling back and forth and lapping back over at the rounded ends. A rocker churn which rocked on a curved base like a cradle was also to be found. The *box churn* was normally small and used by cottagers. Made of wood, it was not necessarily box-shaped (many, in fact, were round) but invariably had slatted paddles inside, usually four-bladed, which were connected to a central spindle, cranked from outside by a handle. The handle, turned vigorously, would revolve the paddles, churning the cream.

The *barrel churn* gradually gained favour from the late eighteenth century. There were two types, one in which the barrel lay on its stand horizontally and revolved longitudinally, and the other in which the barrel sat vertically on its stand and turned end over end. Both, as the name suggests, were constructed in exactly the manner of an ordinary barrel. Both were mounted on sturdy four-legged wooden stands, which were usually fitted with small iron fold-away carrying handles. The horizontal types were the earliest in use. They had an aperture in the side, normally about 9 inches to 1 foot across, into which the cream was poured

ABOVE LEFT: *Coopered plunger churn in use in Wales in the late nineteenth century.* ABOVE RIGHT: *Earthenware plunger churn.* BELOW: *Rocker churn, a nineteenth-century Welsh pattern.*

ABOVE: *Rectangular box churn, late nineteenth century.* BELOW: *Rectangular box churn: internal view showing arrangement of slatted wooden paddles.*
OPPOSITE: *End-over-end barrel churn, late nineteenth century.*

ABOVE LEFT: *Horizontal barrel churn, mid nineteenth century.*
ABOVE RIGHT: *Triangular-type horizontal barrel churn, produced by G. L. Llewellin & Son of Haverfordwest.*

and from which the butter was removed. The small aperture door was normally secured with a screw clasp. This type of churn needed obliquely fixed slats on its inner surface to agitate the cream. The whole barrel was revolved on its stand by cranking a handle at one end, though some were fitted with a handle at both ends. An unusual horizontal churn, which dispensed with the internal slats, was a triangular model produced by G. L. Llewellin & Son of Haverfordwest.

The end-over-end barrel churn became popular towards the end of the nineteenth century. On this type the whole of one end was removable to facilitate filling, emptying and cleansing. This lid was held in place by four metal screw clamps securing it against a rubber seal. It also had a small bung for draining near the base and a glass inspection 'eye' on the lid which allowed the operator to assess the progress of the butter without opening the churn. End-over-end churns needed no internal slats as their tumbling motion was sufficient to make the butter come. Turning on two iron pivots fitted to the balance point of the churn, this type was also normally cranked by handle.

Cottagers at this time also used glass butter jars for making small amounts for personal use. One type was a jar on a pivoted stand which was turned end over end on the principle of the barrel churn. The other was operated by turning a handle which, by means of a ratchet, turned wooden slats inside the jar on the principle of the box churn.

Horses were used on gearings with larger churns, while donkeys and dogs were trained to operate treadwheels to revolve churns, and at least one enterprising farmer fixed a small saddle to the top of a rocker churn filled with cream and encouraged his small son to use it as a rocking horse.

The inside of the churn was washed out with salt water before use to prevent butter from sticking to the sides. When the cream was poured in it was common practice in winter months to add a drop of warm water to help the butter 'come' more quickly. If it failed to come in a reasonable time it was said to have 'gone to sleep'. Constant inspection was necessary to ensure that the butter was not over-churned, as this would impair whey separation. When the butter was just right it was said to have perfect 'grain', which referred to a detectable movement in the

ABOVE: *Rounded box churn, early nineteenth century.*
BELOW: *Glass butter jars: (left) end-over-end; (right) paddle action.*

ABOVE: *Butter churn at the Welsh Folk Museum operated by a dog-powered treadmill. Gearing from the treadmill revolved paddles inside the churn. There are three box-type churns in the background.*
LEFT: *Butterworker on its stand.*
OPPOSITE TOP: *Butter press and pair of butter beaters, used to extract moisture from the butter.*
OPPOSITE CENTRE: *Two sets of butter-board and scotch hands.*
OPPOSITE BOTTOM: *Hand-carved butter stamps.*

butter when it had just come to perfection. The whey was then drained off to be used later as buttermilk or for pig food.

The butter was then washed in cold water and transferred to the *butterworker*, a device for removing excess moisture from the butter. Models varied but most were in the form of a shallow flat-bottomed wooden trough some 3 feet long, 18 inches wide and 3 inches deep. Mounted on a four-legged wooden stand, the trough sloped gently towards one end, where there was a bung hole. Across the trough, secured by tiny wheels fitted in runners on the outside surface, was an iron frame supporting a deeply grooved wooden roller, the spindle of which was connected to a crank handle: the whole resembled the top portion of an old washing mangle.

The butter was spread in the trough and the handle was cranked, causing the whole roller assembly to move up and down the length of the trough on its runners, the grooved roller squeezing out the remaining

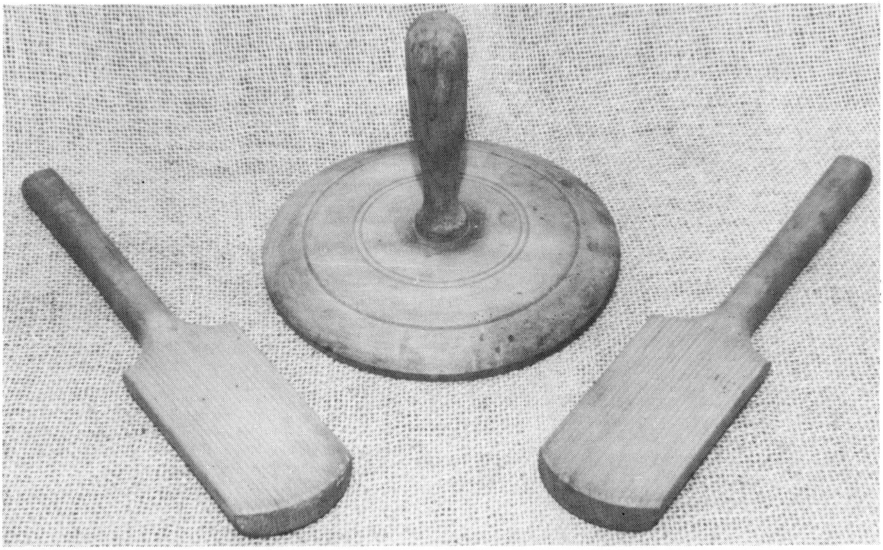

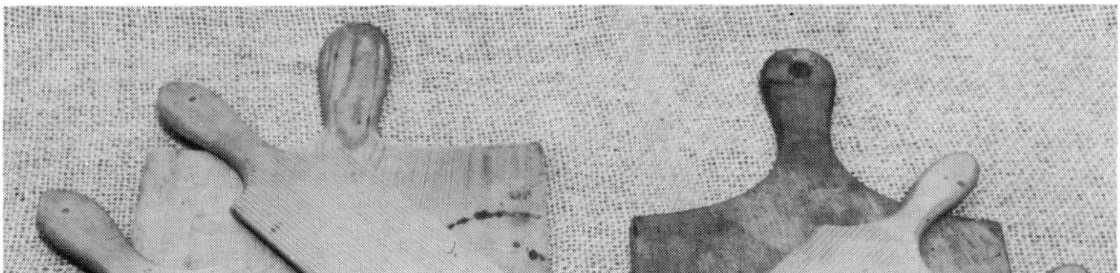

Wooden butter scales.

whey, which then drained off through the bung hole. The operator would all the while be dabbing the butter with a muslin cloth to remove moisture on its upper surface. Some models had trough beds sloping to both ends and roller grooves could run the length or the circumference of the roller. Earlier examples were usually table-top models, the roller being used by hand in the manner of a rolling-pin. Before the advent of butterworkers moisture was removed from butter either by kneading with the hands or by using *butter beaters*. These were usually carved from sycamore or alder and resembled very large, heavy butter pats. They were used to beat and squeeze out the moisture. If required, salt was added by sprinkling on top and working in with a piece of muslin.

The butter was then removed to a flat butterboard and worked into the desired shape with *butter pats* or *scotch hands,* a pair of thin wooden bat-shaped implements, each some 9 inches long, with serrated face edges. The butter, made into rectangular blocks, rounds or rolls, was often finished with an attractive pattern made by dragging the serrated face of the butter pat across it at various angles. Butter was also decorated with a *butter stamp,* a small wooden stamp, normally hand-carved, depicting any one of scores of different designs ranging from flowers, birds and cows to abstract shapes. Some stamps were incorporated into small wooden moulds. There were also wheels which were rolled along the butter leaving an imprinted design and small rolling-pins divided into sections, each carrying a different motif. All butter beaters, butter pats and moulds were washed in salt water before use to prevent the butter adhering to them.

Owing to the poorer quality of grazing winter butter was often paler in colour than summer butter. Salting darkened butter and in Victorian times proprietary brands of an additive called butter colour could be obtained. Cottagers, however, had their own remedies for pale butter. Carrot water and marigold flowers were used quite commonly to bring a richer colour.

Clotted cream was produced in the small dairy by setting milk in tinned cream pans, then heating it slowly over a low flame for an hour or two, making sure it never came to the boil. After being taken from the heat the dish was allowed to stand overnight to allow the crust to thicken and become firm. The scalded cream was skimmed off the following morning with a fleeter.

Unjacketed copper cheese vat.

CHEESEMAKING

Processes for making different types of cheese vary in detail but in most cases follow a similar basic pattern and use the same basic implements. We will consider here the making of the best-known of the cheeses, a farmhouse Cheddar, the standard procedure for which was established by Joseph Harding in 1856. Until then Cheddar cheese had been a hit-and-miss affair with end results varying considerably.

Untainted evening milk was placed in the cheese vat, covered with a blanket to help it ripen and left overnight. The following morning's milk was then added and tested for acidity. If, as was often the case in colder weather, the milk lacked acidity it was often found necessary to add a culture or starter prepared the previous day.

The *cheese vat* came in two distinct types. The earlier was simply a large iron cauldron or coopered wooden tub with a tap at its base; the later was far more sophisticated, being normally a rectangular wooden trough about 4 feet long, 2 feet 6 inches wide and 2 feet deep. This size would contain about 100 gallons. Lining this was a tinned jacket into which hot water was poured, or alternatively cold water heated by steam. The vat had three taps at one end: a large tap for running off the whey and two small taps, one above the other, the bottom one for draining the jacket, the top one as an overflow to prevent the vat jacket from bursting. The

ABOVE: *(from left to right)* Specimen cheese mould and follower. 'Silver Churn' butter-colour jar. Cheese sampler, used to remove some of the core from the cheese for testing. BELOW: Wooden cheese vat with tinned water and steam jacket liner.

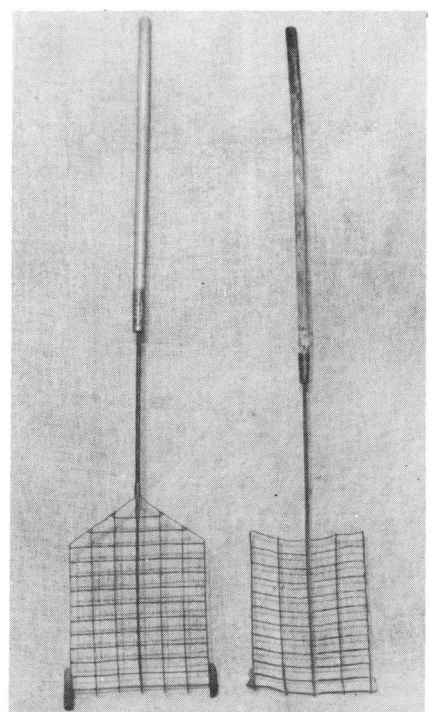 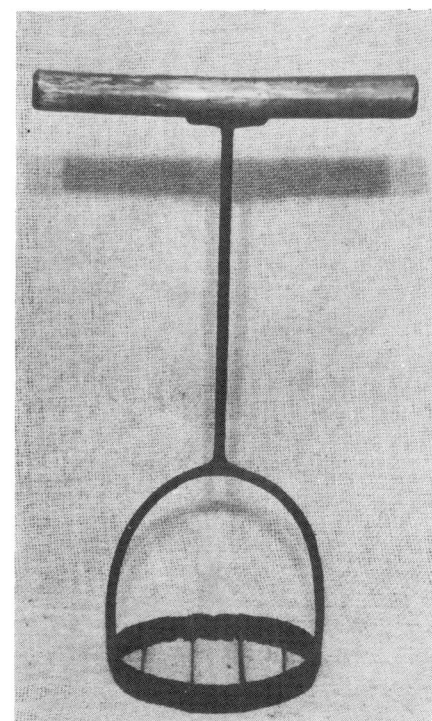

ABOVE LEFT: *Curd agitators.* ABOVE RIGHT: *Early curd knife.*

milk in the vat was then heated. (When using the old unjacketed vat this was done in an old copper boiler over a fire, the milk then being bucketed into the vat.)

The milk was stirred and slowly brought to a temperature of 80 to 82 degrees Fahrenheit. At this point the rennet was added. Rennet normally came in a stoneware jar; it was a foul-smelling reddish-brown liquid prepared from the fourth stomach of suckling calves. These stomachs were known to the cheesemakers as 'vells'. Rennet, stimulated by the acidity in the milk, worked to coagulate it. After the addition of the rennet the milk was stirred, until it started to bubble, with a *curd agitator*. This implement originally took the form of a small two-sided wooden rake but later developed into a shovel-shaped instrument the head of which was a meshwork of brass slats, five or seven thick supporting slats running vertically and upwards of a dozen thinner ones connecting them laterally. A slender brass or iron stem ran from the head to a curved beechwood handle. The curd agitator was usually about 4 feet 6 inches long.

After the curd had been agitated it was left for twenty to twenty-five minutes. The junket (the name for milk coagulated with rennet) should then have been thick enough to cut. This was tested by inserting a finger beneath the crust and raising it. If it broke cleanly it was considered fit. The junket was cut with a *curd knife* about 20 inches long. Earlier examples had wooden cross-piece handles from which ran a narrow iron shaft that forked at its lower end and fitted to an oval of copper or coppered steel that was flattened vertically to form a cutting blade. Across the inside of the oval ran more vertical metal blades. This implement was plunged up and down in the vat to chop the junket to fragments. This device was later replaced by two curd knives used as a pair. Each had a rectangular framework of thin metal cutting strips surmounted by a short wooden

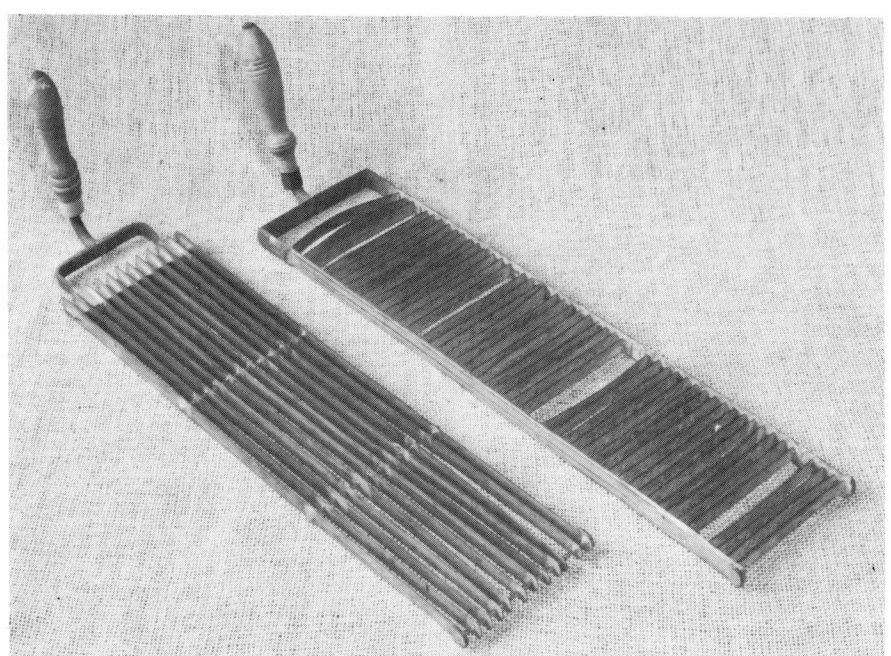

Two Victorian curd knives used as a pair to cut junket, horizontally and vertically, into fragments.

handle fixed at right angles. One knife had its blades arranged horizontally, the other vertically and they were drawn alternatively through the junket. There was also a single curd knife, with half its blades fitted horizontally and half vertically, to do the job of the pair described above.

The next stage was to scald the junket by slowly raising the temperature to between 102 and 104 degrees Fahrenheit. It was stirred again and then left so that the thickened curd would settle to the bottom of the vat and separate from the whey, which was then drained off through the whey tap. The curd was then cut into squares and 'blocked', one square on top of the other, and left to ripen for a time. It was then ground in a *curd mill,* which was normally clamped across the vat. At the top the curd mill had a tapering hopper into which the curd was placed; beneath this was a roller fitted with spiralling rows of metal teeth which, when revolved by a cranked handle, passed through a metal grille, grinding the curd to tiny fragments. The curd was then salted.

The preparation of the cheese was complete: the next stage was the pressing. The chopped curd was placed in a *cheese mould* or *chesset* lined with clean muslin. Early cheese moulds were cylinders formed of wooden slats and bound with iron or brass. In the second half of the nineteenth century these were largely replaced by tinned metal moulds, some with solid bases, others with partly solid bases: most had perforated walls to permit draining of the whey. The table opposite shows the sizes of moulds most commonly used for different varieties of cheese.

After the curd had been placed in the lined mould a wooden block, slightly smaller in diameter, was placed on top. This was called a *follower.* Two or even three of these were often placed on top, depending on the depth to which the curd filled the mould. The mould was then placed on the bed of the press.

Cheese presses varied considerably in design. The earliest were just heavy stones

Variety of cheese	Height (inches)	Diameter (inches)	Weight of cheese (pounds)
Cheddar	15	15	90
Truckle Cheddar	12	7¼	14
Dunlop	15	8	56
Cheshire	15	15	90
Leicester	6	18	38
Derby	6	16	35
Double Gloucester	9	14	30
Single Gloucester	4	15	16
Lancashire	4	15	16
Caerphilly	2¼	10	8
Stilton	12	8	14
Wensleydale (Flat)	5½	10	14
Dorset Blue	4½	10	12

placed on a mould. Later the stone was built into an iron or wooden frame and lowered by an iron thread. This type was superseded by a wooden frame mounted on a heavy wooden bed supported by short wooden legs. The weight of the stone was replaced by a carved wooden block suspended by a flat iron bar perforated at measured intervals. This was fixed by an iron pin to a double-lever system which brought the necessary pressure to bear on the mould when an iron or stone weight was hung on a long extension arm. These were known as *lever presses*. In the Victorian era these made way for lever presses with the frame cast in iron. A metal thread replaced the perforated bar and this enabled the block to be screwed down tightly to the mould. Extra pressure was exerted on the top of the thread column by a short-armed double-lever device operated by relatively small weights hanging on chains at the side. These came as single or double models mounted on cast-iron or wooden beds. Late in the nineteenth century the *spring press* was introduced, which dispensed with levers and weights, the compression of a powerful spring producing the necessary pressure. Smaller spring and screw thread presses were also made at this time for home use.

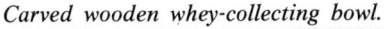

Carved wooden whey-collecting bowl.

LEFT: *A curd mill.*
BELOW: *Detail of the curd mill showing the spiralling arrangement of metal teeth.*

Metal and wooden cheese moulds on a carved wooden cheese bench.

The cheese was left overnight in the press to allow the pressure to squeeze out any remaining superfluous whey. It was then turned in the mould; at this stage the cheese should be firm and have a clean nutty smell. The next step was to place the mould in a bath of warm water to encourage the curds to coalesce. It was pressed again overnight. The next morning the muslin wrap was removed and the cheese was smeared with hot water or melted lard. It was then rewrapped in muslin and a thick linen bandage was sewn round to preserve its shape. It was date-stamped and put to store on the shelves of a cool cheese room.

Many farms which have ceased to make cheese still retain the old cheese rooms, normally long and narrow, with the walls lined with shelves. Many were built partly sunken into the ground to help maintain the even temperature essential to the maturing process of cheese. The caves of Cheddar were used for this purpose for hundreds of years. When in store the cheese was turned every day for two weeks in winter, twice daily in summer. This was done to ensure that the fat globules did not settle at the bottom. After the first fortnight the cheese was turned every other day.

A by-product of cheesemaking was whey butter. The remaining cream was separated from the whey by setting, as with normal buttermaking, then processed similarly. The remaining whey was normally used as pig food.

By about 1920 most cheese was produced in enormous vats in large commercial dairies. Milk was handily dispensed in bottles and, although some farmers still clung doggedly to the long-practised skills of the farmhouse dairy, the days of rural production were numbered. The time-honoured methods, shrewd bucolic wisdom and traditional implements were destined to become bygones.

ABOVE: *Cheese bandage and Cheddar cheese stamp.*

OPPOSITE: *Cheese press of the wooden-lever pattern, with wooden mould, c. 1860.*

FURTHER READING

Fishwick, V.C. *Dairy Farming Theory and Practice.* Crosby and Lockwood, 1947.
Partridge, Michael, *Farm Tools through the Ages.* Osprey, 1973.
Stephens, Henry. *Book of the Farm* (volume II). Blackwoods, 1842.
Walker-Tisdale, C. W. *Practical Cheesemaking.* Headley Bros, 1919.
Wright, Philip. *Old Farm Implements.* A. and C. Black, 1961.

PLACES TO VISIT

Acton Scott Working Farm Museum, Wenlock Lodge, Acton Scott, near Church Stretton, Shropshire. Telephone: Marshbrook (069 46) 306.
Berrington Hall, Leominster, Herefordshire HR6 0DW. Telephone: Leominster (0568) 5721.
Bicton Park Countryside Collection, East Budleigh, Devon. Telephone: Colaton Raleigh (0395) 68465.
Breamore Countryside Museum, Breamore House, Fordingbridge, Hampshire. Telephone: Downton (0725) 22233.
Bygones at Holkham, Holkham Park, Wells-next-the-Sea, Norfolk. Telephone: Fakenham (0328) 710806.
Cambridge and County Folk Museum, 2 and 3 Castle Street, Cambridge. Telephone: Cambridge (0223) 355159.
Cherries Folk Museum, Cherries, Playden, Rye, East Sussex. Telephone: Rye (0797) 3224.
Chewton Cheese Dairy, Chewton Mendip, Bath, Avon. Telephone: Chewton Mendip (076 121) 560.
Cogges Farm Museum, Cogges, Witney, Oxfordshire. Telephone: Witney (0993) 72602.
Cotswold Countryside Collection, Northleach, Gloucestershire. Telephone: (summer) Northleach (0451) 60715, (winter) Cirencester (0285) 5611.
Country Life Museum, Sandy Bay, Exmouth, Devon. Telephone: Exmouth (0395) 274533.
Dairyland Country Life Museum, Tresillian Barton, Summercourt, Newquay, Cornwall. Telephone: Mitchell (087 251) 246.
Dorset County Museum, Dorchester, Dorset. Telephone: Dorchester (0305) 62735.
Elvaston Castle Museum, The Working Estate, Elvaston Castle Country Park, Elvaston, Derbyshire. Telephone: Derby (0332) 71342.
The Great Barn, Avebury, Marlborough, Wiltshire SN8 1RF. Telephone: Avebury (067 23) 555.

Guernsey Folk Museum, Saumarez Park, Catel, Guernsey, Channel Islands. Telephone: Guernsey (0481) 55384.
Ingram-Fowler Country Life Museum, Cricket St Thomas Wildlife Park, near Chard, Somerset. Telephone: Ilchester (0935) 840103.
Manderston, Duns, Berwickshire. Telephone: Duns (0361) 83450.
Mead Mill Collection, Mead Mill, Mill Lane, Romsey, Hampshire SO4 8EQ. Telephone: Romsey (0703) 513444.
Museum of East Anglian Life, Abbots Hall, Stowmarket, Suffolk. Telephone: Stowmarket (0449) 612229.
Museum of English Rural Life, University of Reading, Whiteknights Park, Reading, Berkshire RG6 2AG. Telephone: Reading (0734) 875123 extension 475.
Museum of Lincolnshire Life, Old Barracks, Burton Road, Lincoln LN1 3LY. Telephone: Lincoln (0522) 28448.
National Dairy Museum, Wellington Country Park, Riseley, Reading, Berkshire. Telephone: Heckfield (073 583) 444.
North of England Open Air Museum, Beamish Hall, Beamish, Stanley, County Durham. Telephone: Stanley (0207) 231811.
Old Kiln Agricultural Museum, Reeds Road, Tilford, Farnham, Surrey GU10 2DL. Telephone: Frensham (025 125) 2300.
Ordsall Hall Museum, Taylorson Street, Salford, Manchester M5 3EX. Telephone: 061-872 0251.
Oxfordshire County Museum, Fletcher's House, Woodstock, Oxfordshire OX7 1SN. Telephone: Woodstock (0993) 811456.
Priest's House Museum, 23 High Street, Wimborne Minster, Dorset BH21 1HR. Telephone: Wimborne (0202) 882533.
Rutland County Museum, Catmos Street, Oakham, Leicestershire. Telephone: Oakham (0572) 3654.
Ryedale Folk Museum, Hutton-le-Hole, North Yorkshire YO6 6UA. Telephone: Lastingham (075 15) 367.
St Albans City Museum, Hatfield Road, St Albans, Hertfordshire AL1 3RR. Telephone: St Albans (0727) 56679.
Scolton Manor Museum, Spittal, Haverfordwest, Dyfed. Telephone: Clarbeston (043 782) 328.
Somerset Rural Life Museum, Abbey Farm, Chilkwell Street, Glastonbury, Somerset BA6 8DB. Telephone: Glastonbury (0458) 32903.
Stacey Hill Collection of Industry and Rural Life, Southern Way, Wolverton, Milton Keynes, Buckinghamshire MK12 5EJ. Telephone: Milton Keynes (0908) 316222.
Upminster Tithe Barn Agricultural and Folk Museum, Hall Lane, Upminster, Essex. Telephone: Romford (0708) 44297.
Vale and Downland Museum Centre, Church Street, Wantage, Oxfordshire. Telephone: Wantage (023 57) 66838.
Weald and Downland Open Air Museum, Singleton, Chichester, West Sussex. Telephone: Singleton (024 363) 348.
Welsh Folk Museum, St Fagans, Cardiff, South Glamorgan CF5 6XB. Telephone (0222) 569441.
West Yorkshire Folk Museum, Shibden Hall, Halifax, West Yorkshire HX6 6XG. Telephone: Halifax (0422) 52246.